500 Years
of German Cannon

Krupp's Giant Cannon (35.5 cm) at the Paris World's Fair in 1867. The 50-ton gun was intended for a coastal battery.

Gerhard Taube

Schiffer Military History
Atglen, PA

Cover Pictures

Above: Bronze Bombarde of the 15th century.

Below: "K 5" railroad gun firing, on a Third Reich propaganda postcard.

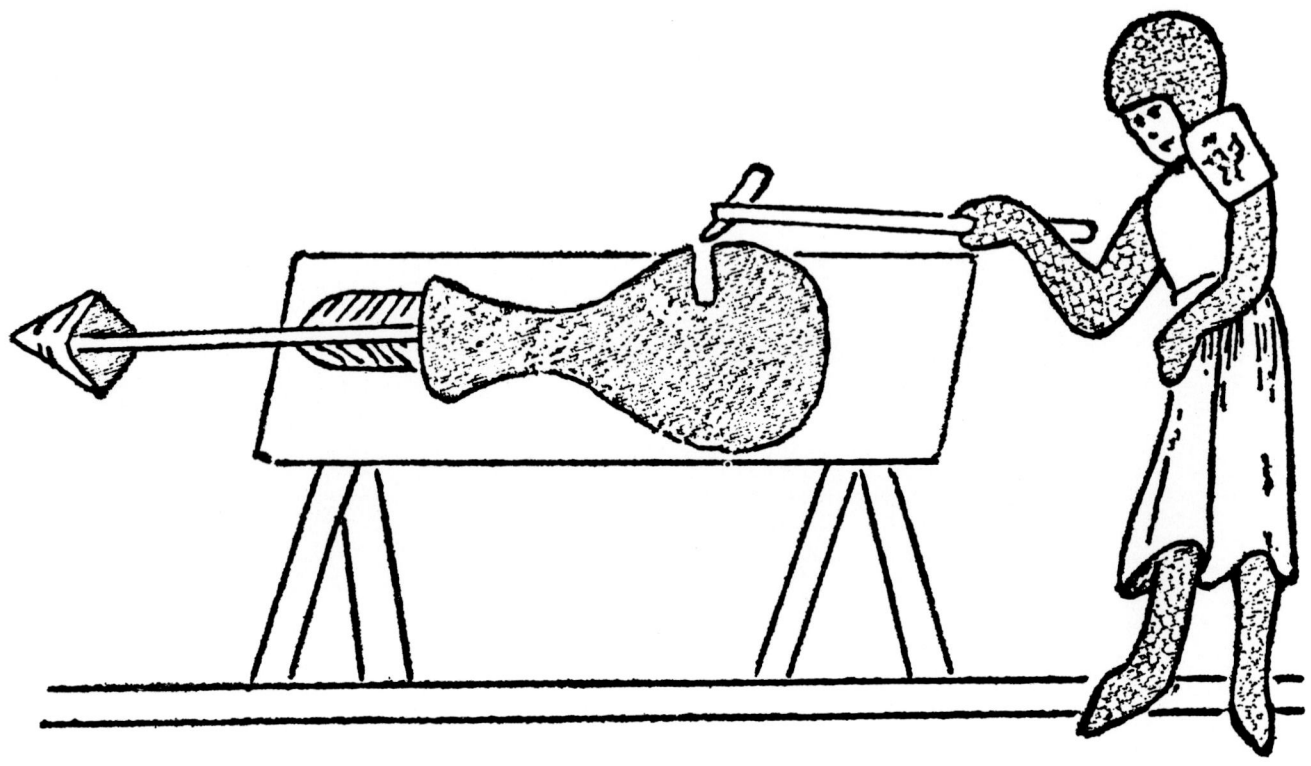

Presumably the oldest portrayal of a European "Bottle Gun," dating from 1326.

Translated from the German by Ed Force

Copyright © 2001 by Schiffer Publishing, Ltd.

All rights reserved. No part of this work may be reproduced or used in any forms or by any means—graphic, electronic or mechanical, including photocopying or information storage and retrieval systems—without written permission from the copyright holder.

Printed in China.
ISBN: 0-7643-1308-8

This book was originally published under the title, *Waffen Arsenal-500 Jahre Deutsche Riesenkanonen* by Podzun-Pallas Verlag, GmbH

We are interested in hearing from authors with book ideas on related topics.

Published by Schiffer Publishing Ltd.
4880 Lower Valley Road
Atglen, PA 19310
Phone: (610) 593-1777
FAX: (610) 593-2002
E-mail: Schifferbk@aol.com
Visit our web site at: www.schifferbooks.com
Please write for a free catalog.
This book may be purchased from the publisher.
Please include $3.95 postage.
Try your bookstore first.

In Europe, Schiffer books are distributed by:
Bushwood Books
6 Marksbury Ave.
Kew Gardens
Surrey TW9 4JF
England
Phone: 44 (0) 20 8392-8585
FAX: 44 (0) 20 8392-9876
E-mail: Bushwd@aol.com
Free postage in the UK. Europe: air mail at cost.
Try your bookstore first.

500 Years of German Cannon

Introduction

History repeats itself very well. The history of weapons in particular is subject to constantly recurring automations. Giant guns, for example, have influenced princes, commanders, and politicians since they appeared in the 15th century. "Ultima ratio regum" (the last argument of rulers) was inscribed on many gun barrels. The muzzle flash of gunpowder weapons was like a form of heat lightning, signaling the beginning of a new epoch in warfare. Thinking in terms of cannons and bunkers soon led to an arms race, a contest between castles and wall-breakers, between gunfire and cement. The high point and end of the "wonder weapon" philosophy in World War II was visibly marked by 60 cm mortars and 80 cm railroad guns—the last hypertrophy of gun-barrel artillery.

This volume offers examples of large-caliber firearms in five centuries, with German guns as examples.

Stone-Boxes

The first known guns in Europe (circa 1326) can be regarded as forerunners. These fat-bellied little objects were called "bottle guns" because of their appearance. From their bronze barrels, some 30 centimeters long, bolts, arrows, gravel, or bits of metal could be fired. In the first half of the 14th century, powder weapons still played a modest role. After that, on the one hand, handguns were developed, and on the other, there appeared the so-called stone-boxes (in regard to their ammunition, namely stones). The stone-boxes were originally cast or forged of iron, and later generally cast in bronze. The outer identifying mark of the stone-boxes is the much smaller powder chamber clearly separated from the barrel itself. With the stone-box, the first actual gun entered the history of artillery. The time was between 1370 and 1380.

Along with the manufacture of stone-boxes, the elite guild of "fire-knowers" or "fire-workers" enjoyed a great advance. The masters of the "new art" jealously guarded their powder recipes and particular techniques and tricks. To be sure, the "unchivalrous weapon" also provoked protests and calumnies. This hindered other contemporaries from improving—or enlarging—the already existing stone-boxes.

Giant Guns in the 15th Century

Toward the end of the 14th century, and at the beginning of the 15th, the gunmakers produced small, medium, heavy, and extremely heavy cannons and mortars. For "calibers" the casters, as the manufacturers were generally called, usually set their own dimensions. Emperor Maximilian I (1493-1519) reduced the variety of calibers in his realm to four main types:

30 cm barrel diameter (caliber) = Haupt- or Steinbüchsen.
24 cm diameter = Scharfmetzen, Nachtigallen (nightingales), or Kartaunen.
5 to 8 cm diameter = Schlangen (snakes).
35 cm and over = Mörser (mortars).

Historians of military technology divide the early guns into three main groups:

Ball diameter approximately 12 to 20 cm = light stone-boxes;
25 to 45 cm = heavy stone-boxes;
50 to about 80 cm = giant guns (including mortars).

The giant guns are regarded by experts as "technically the highest achievements of their times." They attract particular interest from technological historians because "even then, when war was basically the 'ultima ratio regum,' in principle similar demands were made on the resulting development of innovations in weapon technology, such as we know not only from the most recent past, but also from our present-day reality" (Dr. Volker Schmidtchen, military historian).

Production of Bar-Ring Guns

The forged iron stone-boxes were produced all over Europe by essentially the same process. The smiths utilized rectangular iron bars as long as the desired barrel length, arranging them in a ring shape around a wooden form (called a thorn). Over these long rods, red-hot ring-shaped iron bands were placed and "welded" with a hammer at the ends.

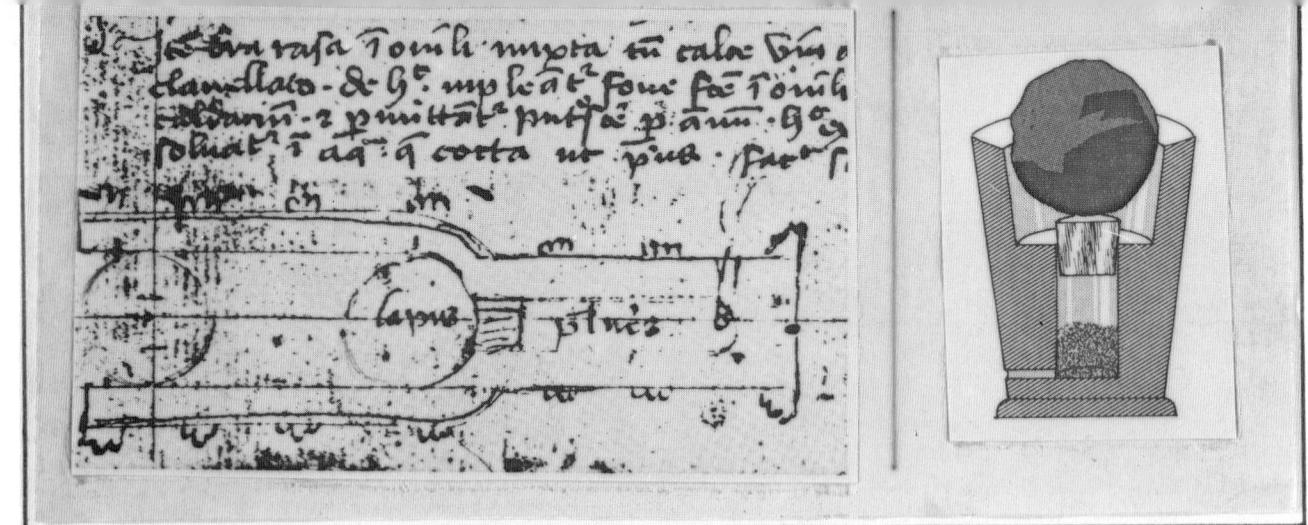

A contemporary pen drawing of a stone-box from 1428 (left) and a diagram of the stone-box principle: powder chamber, wedge-shaped wooden block, stone bullet.

Often the smiths placed several layers of rods and rings over each other, so as to give the barrel great stability. The rings, which contracted as they cooled, pressed the long rods firmly and apparently seamlessly together, giving the impression that they had been welded together. A barrel made this way was permanently under pressure, which stabilized it against the gas pressure from inside when the charge was lit.

The powder chambers, much smaller in diameter, were made separately, usually of "one piece": The "soul" was bored out of a rounded block of iron. The ignition channel was bored from outside afterward. The link between chamber and barrel could be made in two ways. Either the cannonmaker set the chamber in the end of the barrel, welded it to the ends of the rods, and covered the joint with strong iron rings, so as to make the design capable of standing especially great pressure at that point, or the chamber and barrel were first placed in firing position and then screwed together or fastened with wedges that involved the bed of the barrel and chamber. The latter process was favored for especially large and heavy guns, because it made for easier transportation.

A typical stone-box with clearly offset chamber, in this case the "Cologne Stone-box" forged in 1377.

A look into the barrel, approximately 40 cm in diameter, of the "Cologne Stone-box."

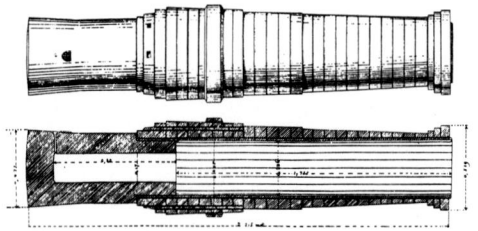

The design of a bar-ring gun.

A large stone-box of the Knights of St. John, sort of a sister gun to the "Lazy Grete of Marienburg," as its shape and dimensions are somewhat similar.

The largest and oldest surviving stone-box made in bar-ring style is the "Pumhart of Steyr" (late 14th-early 15th century); the name "Pumhart" came from "Bombarda" or "Bombarde." The gun is presently in the Military History Museum in Vienna and is often visited for study. Its technical data: Overall length 259 cm, total weight 8 tons, barrel made of 29 long bars 130 cm long and 2 cm thick, plus 12 enclosing rings, over them another layer of 32 iron bars, again enclosed by a layer of rings. The chamber is 115 cm long, the powder charge would weigh about 15 kilograms. The (conical) barrel diameter at the end is 76 cm, at the mouth 88 cm. The stone ball diameter 80 cm, its weight 690 kg. According to calculations by ballistics experts, the "projectile" could be fired about 600 meters with the barrel elevated some ten degrees, with an initial velocity presumably reaching 150 meters per second. The gas pressure was 500 to 600 kg per square centimeter.

"Pumhart of Steyr," with the chamber-barrel connection easy to see. The hooks and eyes made for easier handling of the gigantic stone-box.

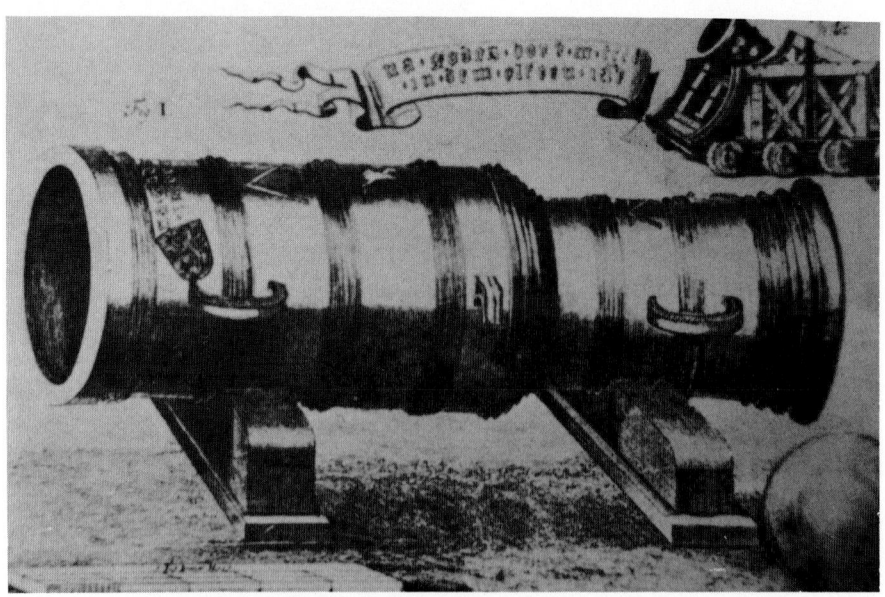

The "Lazy Mette (or "Metze") of Braunschweig," from a 1771 print.

Cast Bronze Guns

The casting of cannon barrels was at first done using the process well known from bell-casting, with the metal poured from above into a form set into the ground. The bronze alloy used for guns usually consisted of 90 parts copper and ten parts tin. Small deviations from the mix percentages or slight additions of lead or iron, as well as the exact moment of their addition, were well-guarded secrets of the master caster. Usually the gun was a single casting, with chamber and barrel in one piece. Acrewed designs were also common: Chamber and barrel were transported separately and only screwed together in firing position. An example is the Turkish "Dardanelles Gun," which proves the proper functioning of such a design beyond a doubt.

The generally bad experiences gained in casting "from above" inspired a different process at the end of the 15th century. The bronze was poured via a casting channel into the form from below. After the principle of communicating barrels, the liquid bronze rose slowly to the top, with air bubbles, slag, and other foreign bodies forced upward into an overflow area. This "lost head" was cut off after cooling.

A German giant gun, the "Lazy Mette of Braun-schweig," in firing position. "Mette" was cast in 1411 and recast into several smaller guns only in 1787.

Ammunition

These giant cannon, used as a rule for breaking walls, usually fired stone balls, some of which were "hardened" with "iron rings" in order to prevent the stone from breaking at firing or at impact with the target. But there were already iron balls, as well. These so-called "full balls" (of cast iron) could also be heated cherry-red before firing. The targets of these glowing balls were timbers, roofs, and wooden fortifications.

Hollow cannonballs also existed since the 14th century. The early explosive shells, like the cartridges used later in the 16th century, were generally used against "soft targets," such as bands of foot soldiers and closed formations. Balls linked by chains (chain-balls) were used mainly at sea against the tackle of enemy ships, Only the classification of guns according to shot weight brought an end to the unregulated variety of guns and ammunition.

Firing a stone-box, this one a so-called "lying piece" with a wooden recoil bed.

Siege Artillery, such as the "Lazy Grete of Marienburg"

Medieval warfare consisted basically of open-field battles and the besieging/defending of fortified places (withdrawal to the "last safe place"). The gunpowder that provided driving energy for wall-breaking ammunition was also to give the impetus for a decisive breakthrough in warfare. Of course, only rich princes and rulers or very affluent cities could afford the new weapon technology. Giant guns were the most expensive part of war materials.

A product of the mighty armaments of the German knightly orders is the renowned "Lazy Grete of Marienburg" (the adjective "lazy" refers only to the slow rate of fire). Within the gun's cost of 1,475 Prussian Marks, the cannon alone cost 278 Marks—then a tremendous expenditure for a single gun. The total length was 250 cm, the ball diameter 50 cm. The weight of the ball was some 170 kilograms. The total weight of the cannon was almost five tons. To transport it, depending on the terrain conditions, up to 25 horses were

A heavy stone-box with a very long wooden recoil bed, which was weighted down with stones or earth.

required. Moving the other equipment, such as the cradle, ammunition, and tools, required a train of some thirty wagons and 150 draft animals. This medieval artillery unit saw very successful service in the Mark Brandenburg in 1413.

There the Burggrave of Nuremberg, the later Elector Friedrich I of Hohenzollern, who had been named viceroy by Emperor Sigismund in 1412, was having very serious trouble with the Markish robber-barons led by the brothers Johann and Dietrich von Quitzow. The imperiled Friedrich saw hope only through the use of the knightly order's giant gun. Early in 1413 the Teutonic Knights turned their remarkable gun, along with its crew, over to the Burggrave in Kuestrin. The Quitzows, behind their four-meter walls, calmly watched their enemies approach. Along with its smaller sister guns, the "Lazy Grete" reduced the four chief castles to a stormable state in the time from February 9 to 26, 1413. In barely three weeks, well-used firearms laid the cornerstone for the rise of Brandenburg-Prussia.

Fortress Artillery

Like every new weapon, the siege guns inspired the development of opposing weapons. Along with construction changes, such as reinforcements, omission of wooden buildings, and creating higher defensive walls and bastions, special fortress artillery was also developed. These guns were characterized by considerably smaller calibers, much longer barrels, greater range, and higher targeting accuracy. In particular, the long-barreled cannons, or "Basilisks," developed at the beginning of the 16th century became a sort of standard gun for fortress artillery. Even then it was not a case of every gun crew firing at will. There was already a central fire-control command, either the fortress commandant himself or his division commanders took over that command.

The muzzle of a bar-ring gun of the 15th century. Only remains of the long bars can be seen, but the rings of the outer mantle clearly show the construction principle.

Above: The "Lazy Maid of Dresden" (1420). This heavy stone-box (bar-ring gun) has an overall length of 235 cm and a bore diameter of 34 cm; the ball weight was 53 kilograms. The total weight was 1.3 tons. The wheeled carriage was built only around 1500.

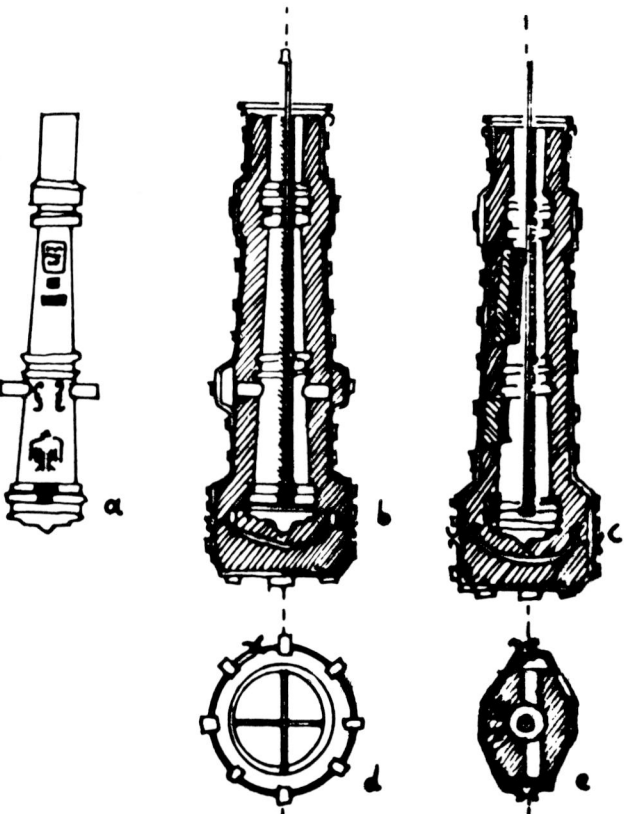

Above right: Cast bronze guns:
a. Completed gun barrel with overflow.
b. Cutaway showing casting shape in the plane of the trunnions.
c. Cutaway showing casting shape at the trunnion plane.
[needs clarifying]
d. View of the inner base form (from below).
e. Cutaway showing casting shape at trunnion height.

Below: A gun foundry circa 1570. In the upper right background is a siege scene.

Above: Siege guns of the 15th century: large "boxes," a mortar and a multiple gun.

Right: A breech-loading gun at Marksburg castle (16th century), and a sketch of its functioning.

A "Dardanelles Gun." The screw design is a superb achievement of the Turkish builder (1469). Overall length 518 cm, chamber length 215 cm, ball diameter 63 cm, ball weight 340 kg. This bronze stone-box weighs 18.6 tons.

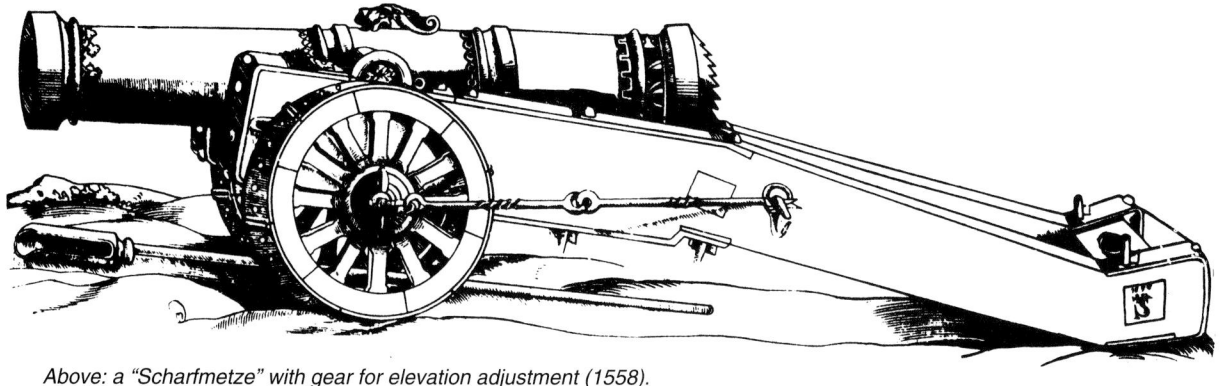

Above: a "Scharfmetze" with gear for elevation adjustment (1558).

An early breech-loader, circa 1460-70, which was not very successful on account of the often unsealed chamber closing.

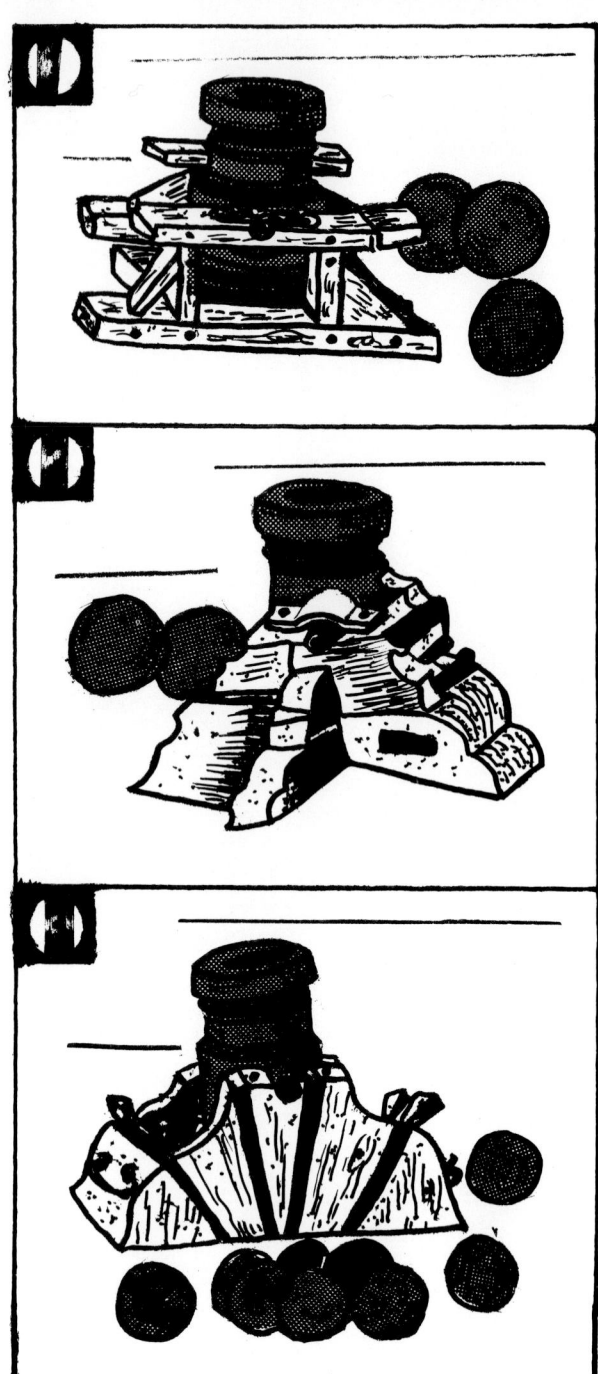

Left: Elevated fire:
1 and 2: Heavy bronze mortars in firing position (end of the 15th century).
3. Bronze mortar with trunnions; so-called hanging mortar in firing position, with adjustable elevation by wooden wedges (ca. 1500).

Right: Firing and targeting:
a. Measuring the barrel elevation with a quadrant.
b. Aiming with a direction rod.
c. Firing on round and angular towers.

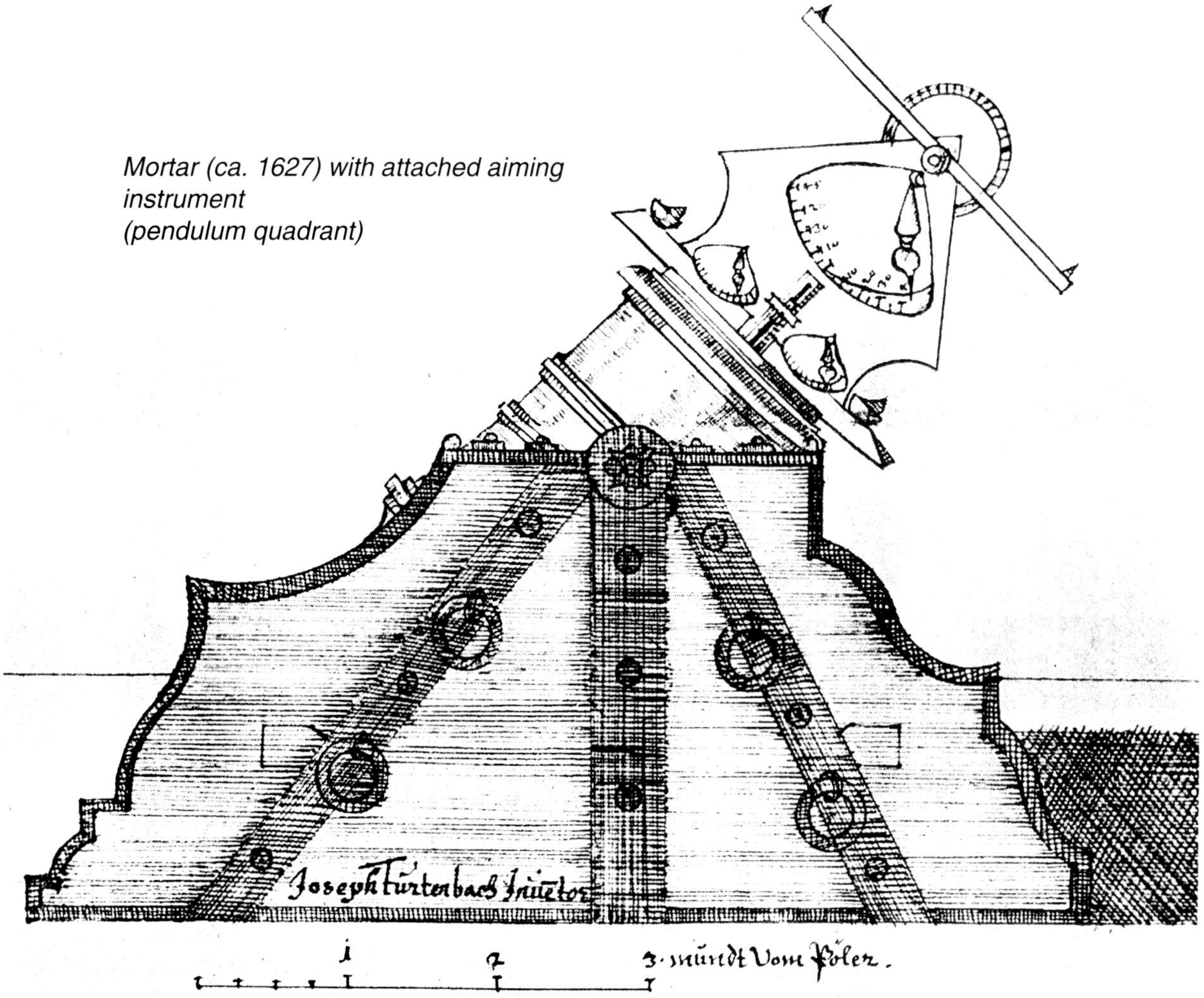

Mortar (ca. 1627) with attached aiming instrument (pendulum quadrant)

Right: The muzzle of the "Griffin" camera (seen here at Ehrenbreitstein fortress).

Competition

The answer to the long-barreled weapons of the fortress artillery did not take long to arrive. In the course of the 16th century, gun barrels were mounted on wheels for better mobility, and iron cannonballs were introduced for better breaking power. The changes in heavy siege guns also became evident in their external appearance (corresponding to the concept of the modern cannon). A good example is the "Greif" (Griffin), a cannon cast in 1524 for the Archbishop of Trier. The Griffin can be seen today at Ehrenbreitstein fortress opposite Koblenz.

Now fortress construction was modified again, systematized according to mathematical, physical, geometrical, and ballistic laws. One of the many answers consisted of mortars, short-bore weapons with steep trajectories (generally over 45 degrees).

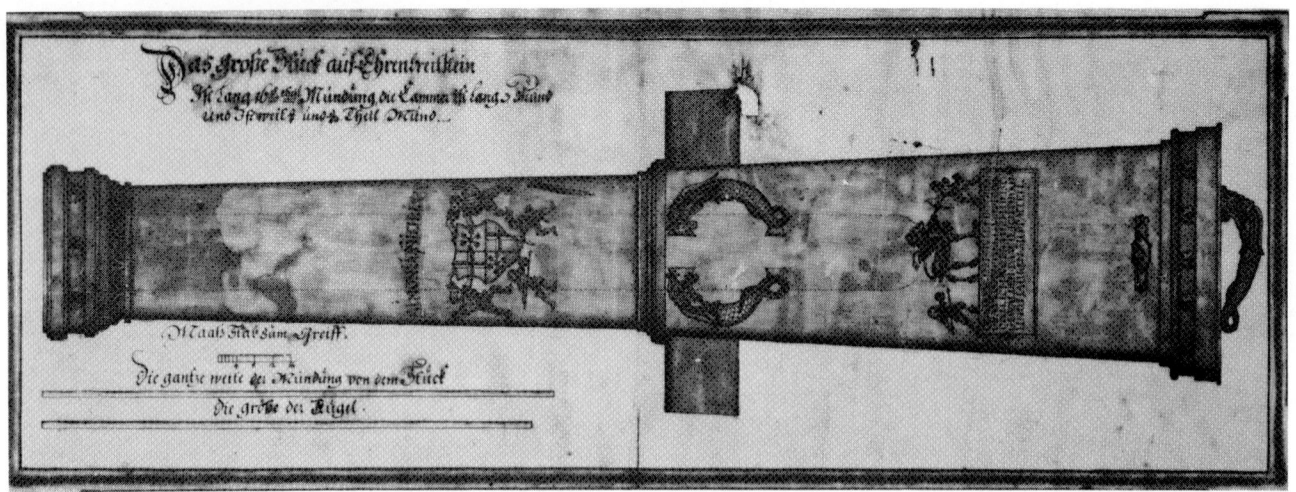

The "Griffin" cannon cast in Frankfurt on the Main in 1524. Length 4.94 meters, weight 9 tons. According to recent tests, the gun could not have been put to use, as the touchhole has not been bored through to the chamber. Experts presume that the "Griffin" was only intended for show.

In the latter half of the 17th century there was something like a blossoming of fortress war. The absolutist princes used fortresses that were linked to form chains of defense to protect their boundaries. For example, the French engineer Marshal Vauban stood out in fortress construction. His principles of fortress building and fighting remained valid into the 19th and in part into the 20th century.

The giant guns of the Middle Ages doubtless incorporated fine achievements in manufacturing. But the mistaken notion that ever-larger guns absolutely had to bring ever-greater victories finally led to a dead end in weapons technology. The giant guns lost significance. Light and mobile field and infantry guns, mass-produced by industry, prevailed. Only the development of the breech-loader, the rifled barrel, and long shells (grenades), plus the invention of the barrel recoil brought with them the rebirth of large-caliber high-angle and long-range flat-trajectory guns, with everything much better perfected.

Famous German Giant Guns

Name	Time period	Material and type	Barrel length in cm	Weight in kg	Caliber in cm
"Lazy Grete of Marienburg"	ca. 1408	bronze cast	250	4600	50
"Lazy Maid of Dresden"	ca. 1410	iron bar-ring	233	1320	34
"Lazy Mette of Braunschweig"	ca. 1411	bronze cast	290	8228	67

Famous foreign guns (for comparison)

Name	Time period	Material and type	Barrel length in cm	Weight in kg	Caliber in cm
"Pumhart of Steyr"	end 14th cent.	iron bar-ring	259	8000	80 (middle)
"Dulle Griet of Ghent"	ca. 1435	iron bar-ring	500	16,400	64
"Mons Meg" (Edinburgh)	ca. 1455	iron bar-ring	400	5790	50

A 30.5 cm coastal battery of the Imperial Navy at Helgoland.

Below: For centuries, big guns influenced princes and politicians. Here a Chinese study commission is visiting Krupp in 1910.

The railroad artillery began with improvisations: 17 cm "railroad cannon" on wheeled mount, set on rollers.

High-Trajectory and Railroad Guns

In Germany, the name Krupp was to become synonymous with high-performance guns. The firm first entered the history of armaments in 1867 as a manufacturer of cannons, making a name for itself at the Paris World's Fair with what was then the biggest gun in the world. The rifled breech-loader was made of high-quality Krupp crucible steel. The 35.5 cm cannon weighed about 50 tons and was intended for use at a coastal fort. These very heavy guns could be used effectively only as fixed-position coastal guns or on battleships and heavy cruisers.

At the beginning of World War I, the following scene existed "on land": France was superior with its heavy, long-range flat-trajectory fire (naval guns in railroad mounts). On the other hand, the German mortars (high-angle guns) proved to be better developed and had more penetrating power than their French counterparts. The Germans also met the French advantage in heavy flat-trajectory fire with naval guns: From 1916 on, the first German railroad guns were rolling to the front.

A 30.5 cm mortar barrel being transported by rail.

Successful German high-angle fire: aiming a 21 cm mortar (World War I).

The arms race between guns and concrete: The Krupp 42 cm high-angle gun (so-called Gamma Device) on the firm's firing range in Meppen.

The 42 cm high-angle "Big Bertha" gun (operated in training).

"Big Bertha"

The miraculous secret weapon called "Big Bertha," a 42 cm Krupp high-angle gun, grew out of the competition between guns and bunkers on the one hand, and the close cooperation between the Artillery Testing Commission (APK) of Imperial Germany and the armaments industry on the other. In 1903-04 the General Staff and the APK requested designs from Krupp for a heavy high-angle gun that could successfully attack the French Gelopin towers with their 30 cm nickel steel and almost three-meter-thick concrete coverings.

The 42 cm high-angle gun "Big Bertha" from another perspective.

Above: The "Becker" Gamma battery in Belgium in 1914.

Building on their good experiences with heavy guns in wheeled mounts (among others, a 30.5 cm mortar, the so-called Beta Device), Krupp developed and built a 42 cm mortar, officially the M-Device. The M stood for mine-launcher, as it was originally intended. For reasons of secrecy it was left with that designation. This mortar on a wheeled mount was the real "Big Bertha," as the people and the patriotic press nicknamed it after its initial successes on the western front. In the literature, the most varying information about "Big Bertha" (guns on wheeled mounts) appears. Here are some data from original Krupp sources: "Barrel weight (all in tons) 13.4, barrel wagon including belt 19.9, mount vehicle with limber 18.3, brace wagon 21, equipment wagon, loaded, 17.6, maximum axle pressure 10.6. Massive barrel, sliding-wedge breech. Wraparound gun shield. Elevation limits 0 to 65 degrees. Shot range 9,300 meters. Long shell L/3.6 without middle plate with 144 kg explosive charge. Since November 1916, also short M-shells with hoods. Battery transport by steam tractors and Podeus tractors."

A 42 mm shell for "Big Bertha (seen at the Imperial War Museum, London).

19

A 30.5 cm Krupp howitzer, designated heavy Kartaune. Weight in firing position 47 tons, maximum range 16.5 km. The gun was supposed to increase the range of heavy high-angle guns decisively.

The gun became famous above all for its initial success, when its cement-breaking shells shattered the stone walls of forts and pulverized brick fortresses. In particular, the "Wesener" mobile battery took part in the fighting around the forts at Liege in August 1914 and, by October, had smashed ten forts through precise and tactically clever firing. Where it was less a question of fast striking, the relatively immobile Gamma Devices (fixed-bed guns) could show their effectiveness. The two Gamma batteries "Becker" and "Solf" were used in Belgium and Lorraine. Later, as in the battle of Verdun, the 42 cm guns no longer played a decisive role.

The "Karl" device with an armored ammunition carrier. Here Japanese officers are inspecting the "Baldur" gun at the test firing range in Hillersleben.

An ammunition tractor on a Culemeyer vehicle, with a tractor in the background.

The "Karl" Device

All the same, the high-angle gun as such took new form in the Third Reich as the "Karl" 60 cm mortar, on a fully tracked self-propelled mount. Along with two mounts, six guns were produced by Rheinmetall and put through firing tests for acceptance by August 1941. Along with the "Karl Device" designation, there was also the "Device 040" camouflage designation for the 60 cm barrel and Device 041 for the gun fitted with a 54 cm barrel after 1943. The troops gave the guns their own names: Adam, Eve, Loki, Odin, Thor, and Ziu. Toward the end of the war the Adam and Eve guns were renamed Baldur and Wotan.

The "Karl Device" loaded for rail transport.

21

This series of pictures shows how the gun was put into firing position.

The upper mount with the gun cradle was moved from the Culemeyer vehicle to the self-propelled mount by means of a crane.

In the same manner, the barrel with its breech followed.

Installing the recuperator completed the gun assembly as far as the main components were concerned.

With the help of a crane, a shell was lifted out of the ammunition tractor...

This series of pictures shows how the "Karl Device" was loaded.

...and swung to the gun.

The shell was lubricated and pushed into the barrel with the rammer.

The 124-ton gun fired heavy anti-concrete shells (2,200 kg, range 4,300 meters) or light anti-concrete shells (1,700 kg, range 6,700 meters). A characteristic of the shells was that the propellant charge was also kept in the shell (base), because the shortness of the gun barrel did not allow the insertion of an older-type shell. In firing position, the device was lowered with its entire base on the ground. It was transported over long distances on special railroad cars. For short distances the gun was transported in separate loads on Culemeyer vehicles pulled by halftracked tractors.

Originally intended for action against the Maginot Line, these guns saw action mainly in Brest-Litovsk (1941), Sevastopol (1942), and Warsaw (1944).

The barrel is placed in the ordered position.

Right: Relics of the heaviest German high-angle fire, seen at the Rheinmetall premises in Düsseldorf: the 60 cm anti-concrete shells with partly rusted ballistic hood (left), and the 54 cm anti-concrete shell without a ballistic hood.

The "Karl Device" in firing position before Sevastopol.

The "Thor Device" after refitting with a 54 cm barrel.

The observer of the 2./833 photographed the citadel of Brest-Litovsk through his shear telescope.

Above: A 21 cm mortar of Heavy Artillery Unit 833 before Sevastopol.

The high-angle "Thor Device" firing before Sevastopol (photo series from "Deutsche Wochenschau").

What the pictures have in common: Above, a crater of a 42 cm shell from "Big Bertha" near Verdun (World War I), and below, a crater of a 60 cm shell from "Karl" in the Lemberg area (World War II).

The 24 cm cannon on the "Karl" mount. The "Karl" cannon project could not be realized.

Above: A makeshift 15 cm railroad cannon from the "Sofort Program." (sofort = immediately)

Above: In the battle for the fortress of Sevastopol, the German command ordered all available guns brought together; here a Krupp 28 cm howitzer is seen firing in a newsreel excerpt.

Below: The 21 cm Cannon 38. This 25.3-ton gun had double recoil fittings. Because of lack of interest from the army command, only eight of these guns were made by 1943.

From 1916 on, heavy railroad guns with flat-trajectory fire were also available. This is the 38 cm railroad cannon "Max."

The "Paris Gun"

As the American government then evaluated it, the 21 cm long-range gun was primarily a psychological and political weapon. The highest shot ranges achieved by naval guns (on railroad mounts) during World War I were around 60 kilometers. Compelled by the war situation, the Army High Command demanded ranges of 120 km! In the most unusual way, Professor Dr. Fritz Rausenberger, director and chief designer at Krupp, was able to fulfill this demand.

During test firing at Altenwalde on January 30, 1918, Rausenberger's team attained, at 50-degree elevation and with 4,300 atm. of gas pressure and a muzzle velocity of 1,645 meters per second, a range of 126 kilometers! The flight time was 182 seconds, the angle of fall 55 degrees. The basic device was the 38 cm railroad gun "Max." In its 17-meter barrel a 21 cm inner barrel had been installed, projecting some 13 meters out of the outer barrel. The raw design was lengthened with a so-called tube six meters long. The total barrel length was thus 36 meters. A support system prevented the overly long barrel from sagging.

The "Paris Gun," lowered onto a turntable mount. In this drawing, the support system has been drawn somewhat bigger than to scale.

The "Paris Gun" at the naval firing range in Altenwalde. The barrel is mounted in a test-firing mount.

The gun was transported to its firing position by rail and then set, by means of a crane, on a specially designed machine-driven turntable mount.

In the period from early March to mid-August 1918, three guns in all fired on Paris from distances of 128, 107, and 87 kilometers. As to the shape of the trajectory, mostly in almost airless space at 40-km altitudes, the round shape of the earth played a role, as did known ballistic characteristics. Because of the long flight time of the shells (up to three minutes), aiming required that the earth's rotation be considered. The 106 kg shell was light in comparison to the very heavy gun (140 tons in firing position). The small explosive charge of 7 to 8 kilograms indicates the mainly psychological aspect of this "miracle weapon."

Railroad Gun "K 12"

The "cannon thinking" of a Hitler also brought about the new edition of the "Paris Gun," the 21 cm Cannon 12 in a railroad mount, known as "K 12" for short. With the "K 12" the Army had the gun with the longest range in World War II. In comparison with its World War I model, the parallels (those of K 12 in parentheses) can be seen: Caliber 21 cm (21 cm), inner barrel length 28 m (32 m), elevation range to +55 degrees (+55), maximum gas pressure 4,800 kp/sq.cm (4,600-5,000), shell weight 106 kg (107.5 kg), explosive charge 7-8 kg (7.85-8.10 kg), loading weight 200 kg (250 kg), muzzle velocity 1,645 m/sec (1,625 m/sec), maximum range 130 km (115 km), ignition (head and base (head and base), loading space 4 meters (4 meters).

Although the "K 12" was the most superbly developed long-range gun, it could not fulfill its expectations—the role of long-range warfare had long since been taken over by the Luftwaffe. The two completed specimens fired from the French coast at Dover, Folkestone, and portions of the County of Kent from late 1940 to 1941 without noteworthy effects.

The actual "value" of this unusual artillery weapon lay in the know-how, as one would say today. The understanding and experience of Rausenberger that influenced the gun, it has been judged, cannot be evaluated in dollars and cents. The technical knowledge gained from the two long-range guns in the two World Wars also affected the "K 5," the standard gun of the German railroad artillery.

Right: A drawing of the "Paris gun" in firing position.

The remains of the turntable were all that the French found of the long-range gun after the German withdrawal.

Right: The new edition of the "Paris Gun": the "K 12" railroad gun, seen here in a typical propaganda photo.

Below: The shell and propellant for firing on Paris.

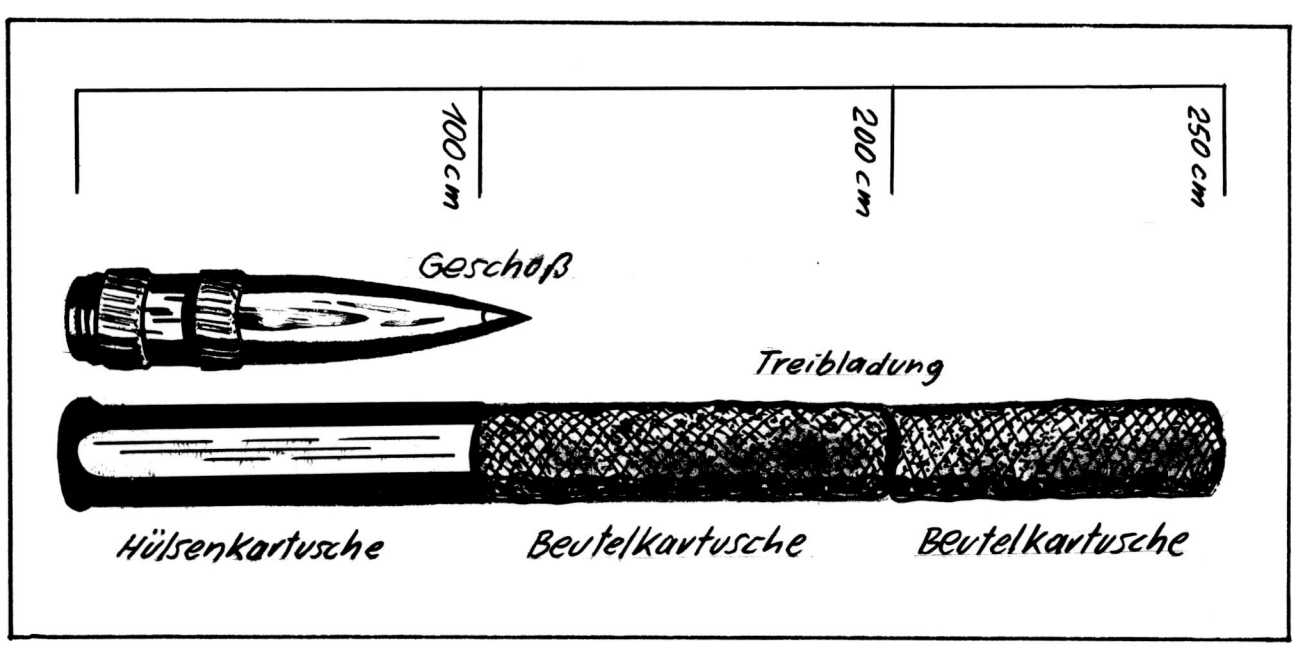

"K 12" near Calais.

The "K 5" Railroad Gun

Behind the brief "K 5" name, synonymous with highly developed weapons technology, there lurked a 28 cm cannon that is regarded as a high point of big gun development, as well as series production (27 were made). The 28 cm Cannon 5 in railroad mount (28 cm K 5 E)—official designation—was the standard gun that formed the backbone of the German railroad artilery. The "K 5" appeared on all major theaters of war.

Its weight in firing position was 218 tons. The guns were fitted with either the so-called deep-rifling barrels (10 and 7 mm) or, later, the multi-rifling barrels, or even smooth barrels (K 5 Glatt, 31 cm). The maximum range with a large charge and a muzzle velocity of 1.128 m/sec was about 52.2 km. As part of the development of special ammunition, which was carried on for almost all large-caliber guns during the entire war on account of permanent requests for increased ranges, the "K 5" guns also received shells with

A rare long-distance photo: Radar antennas near Dover under fire from German long-range artillery.

Above: A "K 5" of Battery 710 firing near Calais in 1941.

rocket propulsion (Rocket Shell 4331). The ranges reached with them were about 86 km. Some guns were equipped to fire the "Peenemünde Arrow Shell." These wing-stabilized undercaliber projectiles (12 cm) were fired from smooth-bore guns (K 5 Glatt) bored out to 31 cm. In test firing at the Army's artillery firing range at Rügenwalde in Pomerania, the highest shot ranges were some 150 km!

Above: A "K 5" on a turntable.

Left: A bunker for railroad guns in Calais.

Above: Battery 710 being transported to the eastern front in 1941 (transport and living wagon).

Battery 710 on the eastern front in 1941.

Below: A "K 5" of Battery 710 with a clearly recognizable deep-rifled barrel.

Below: A Battery 710 camera "position" for velocity measuring, seen on the eastern front in June 1941.

36

A "K 5" (nicknamed "Leopold") captured by the Americans on the Anzio front in 1944, on display at the U.S. Army Museum in Aberdeen (power source and ammunition crane are missing).

Two "K 5" guns fell unharmed into American hands on the Anzio front in Italy in 1944. One is now at the U.S. Army Museum in Aberdeen, Maryland. It is the only German railroad gun still in existence anywhere, in terms of maintenance and care. The Sinsheim Technical Museum has been trying for some time to obtain a 28 cm railroad gun from the "Bruno" family, which exists in France in very poor condition.

The largest caliber of railroad artillery still in one piece (Transit-Profile): the 38 cm "Siegfried" railroad gun (showing the elevation gear here).

The 38 cm railroad gun "Siegfried" with its barrel at 45-degree elevation.

The "Siegfried" Cannon

The 38 cm Cannon in railroad mount "Siegfried" (38 cm Siegfried K E), which resulted from the Army's request for heavy long-range calibers, took a special position within the railroad artillery to the extent that it was the largest-caliber gun that could be transported "in one piece" (Transit-Profile).

The three guns that reached the troops saw service mainly in France for firing on British sea traffic. The weight in firing position was 294 tons, the maximum range was 56 km. The so-called "Siegfried Grenade," weighing 495 kg, was fired. The "Special Siegfried Charge" was divided between the main cartridge (133 kg) and the primer (123 kg).

A 40.6 km cannon in the "Adolf" railroad mount has been written about but never existed. The work on it was soon halted. "Adolf" cannons, though, saw service as fixed coastal guns on the Atlantic Wall and in Norway.

The power source of "Siegfried."

Below: "Siegfried" close-up: The breechblock and shell catcher.

Special Devices "Doria" and "Heavy Gus 2"

The largest and most technically spectacular gun of all times was the 80 cm railroad cannon nicknamed "Dora," officially just called "Special Device" or "D Device." The similar gun "Heavy Gustav II" (or "2nd Device") was almost identical to "Dora," but did not see service at the front. The 500-man Heavy Artillery Unit (E) 672 was established (January 1942) for "Dora."

The "Monster" (so called by Albert Speer) was originally ordered by Hitler personally and was intended for breaking down the French Maginot Line. Since Krupp could not deliver it on time, it was ordered that the giant gun of the 20th century be used against the greatest land and sea fortress: Sevastopol. In logistical terms this was a questionable undertaking. For this purpose a suitable firing position had to be located, built up, and secured in enemy territory. The gun, weighing 1,350 tons and moved on 80 wheels, as high as a three-story house, was almost 50 meters long and 7 meters wide (it moved on double rails), had to be dismantled for transport, and moved in pieces on five special trains with over 100 cars.

Right, below, and opposite page: Assembling the 80 cm railroad gun.

Above: The right side of the mount with coupling gear to link it to the left side.

The gun was usually assembled by two Ardelt portal cranes.

Above: The outer barrel is being mounted.

The inner barrel is installed.

Mounting the barrel: Here the outer barrel is still on its railroad car.

Right: the 80 cm railroad gun "Heavy Gustav 2" at the Army's test firing range in Rügenwalde, Pomerania, in 1943.

Below: A look at the fully assembled gun, with a double Diesel locomotive to move the gun to the firing line.

Above: Arrival near Sevastopol: Ammunition (here a bag cartridge) is being unloaded.

Below: The two-ton propellant charge (one case and two bag cartridges).

The electric ammunition lift, with shells at left, propellants at right.

In June 1942, "Dora" fired 48 armor-piercing shells on five firing days against targets at the fortress of Sevastopol, as ordered by the highest authorities. The seven-ton antitank shells penetrated a meter of steel, 8 meters of reinforced concrete, or 32 meters of soil with plant growth. Only five shots could be scored as hits. Still, this is remarkable when one considers that the targets with their few 100-meter stretches of lateral length amounted to tiny pinpoint targets—at ranges of 25 to 35 kilometers. Five explosive shells, delivered after the fall of the fortress and city of Sevastopol, were fired by the unit for testing purposes. It took no military knowledge to recognize that the gigantic organizational, material, logistic and, not least, personal expenditures were in no way related to "usefulness." The troops were also opposed to these "dinosaur guns" from the start. The artillerymen would much rather have had a large number of smaller, handier guns, particularly the 21 cm Mortar 18, for use in the east. Undeniable, though, was the achievement of the designers, who knew how to turn extraordinary political demands into astonishing results. The tactical failure should not be blamed on them.

Before Sevastopol the 80 cm supergun experienced its first and last "baptism of fire." Later attempts to modify the two artillery giants for firing on England were given up, at the latest in November 1944. A few days before the war ended, German special commandos blew up the artillery parts of "Dora" and "Heavy Gustav 2" in Chemnitz and Grafenwöhr. The rolling components, especially the turntables, were used in part in the civilian sector. A turntable of "Dora," for example, is still to be seen today in the scrap heap of a gravel company in Hedemünden, pushed off onto the unloading sidetrack of the history of German giant guns. Relics such as cartridges and grenades are found in technical and military museums inside and outside Germany.

Right: "Dora" in firing position near Sevastopol.

Lower left: A gun ready to fire, with insulated workshop and ammunition wagons in the foreground.

Lower right: "Dora" firing before Sevastopol.

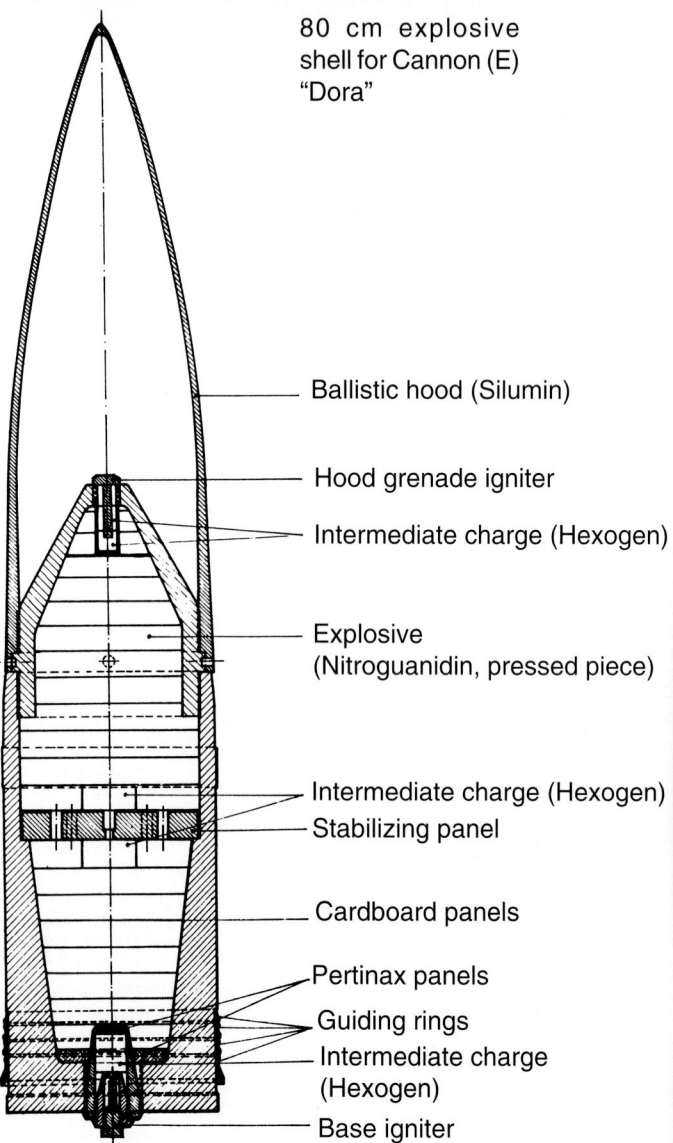

Above: Hauptmann Otto Schmidt (7th Echelon/St.G. 77) photographed the "Dora" firing position before Sevastopol from his Ju 87 plane.

80 cm explosive shell for Cannon (E) "Dora"

- Ballistic hood (Silumin)
- Hood grenade igniter
- Intermediate charge (Hexogen)
- Explosive (Nitroguanidin, pressed piece)
- Intermediate charge (Hexogen)
- Stabilizing panel
- Cardboard panels
- Pertinax panels
- Guiding rings
- Intermediate charge (Hexogen)
- Base igniter

Left: The 80 cm explosive shell. Drawn after being defused by experts in the ex-GDR. The shell was found in a destroyed gun mount in Hillersleben. When it was defused, it was found that the explosive shell—contrary to what had been printed about it—had two fuses, a head and a base fuse.

Below: An 80 cm shell with cartridge, seen in an ammunition dump in Schleswig-Holstein after the German surrender.

Chronology and important dates for "Dora"

1936	OKH contract to Krupp to develop a railroad gun with hitherto unattained penetrating power.
1941-42	Testing and shot-table firing at the army test firing ranges in Hilelrsleben and Rügenwalde.
March-April 1942	Preparations for action in the east, especially training and drill in assmebling and disassembling the gun.
June 1942	Action before Sevastopol.
March 1943	Display and firing of "Dora" before Hitler and many people from industry and Wehrmacht.
April 1945	Destruction by German explosives commandos.

Data

Caliber	80 cm
Weight ready to fire	1,350 tons
Total barrel length	32.48 meters
Total barrel weight	400 rons
Length of gun	47.97 meters
Width of gun	7.1 meters
Height of gun	11.6 meters
Number of turntables	8
Axles per turntable	5
Diesel-electric locomotive power to move the gun	1 x 1,000 HP
Barrel design	Outer and inner barrels

Ammunition (antitank shell)

Weight	7.1 tons
Length without hood	2.4 meters
Length of hood	1.53 meters
Type of fuse	Base
Max. shot range	38 km
Explosive charge weight	250 kg

Penetrating power

Steel	1 meter
Reinforced concrete	8 meters
Soil with plant growth	32 meters

An 80 cm shell (blind) on the destroyed testing mount at the Hillersleben testing range (summer 1945).

The Heavy Flak Guns 1933-1945
Werner Muller
Superb study of the 88, 105, 128 and 150mm guns.
Size: 7" x 10"
140 pp.
ISBN: 0-88740-263-1
hard cover
$24.95

German Heavy Field Artillery in World War II
German heavy artillery as used on all fronts and with a variety of sizes and capabilities.
Size: 8 1/4" x 11 3/4"
b/w photos, line drawings
48 pp.
ISBN: 0-88740-759-5
soft cover
$9.95